Copyright © 2020 by Tom Nicholas

Cover © 2020 Canva

All rights reserved. This book or any portion thereof may not be reproduced or used in any manner whatsoever without the express written permission of the publisher except for the use of brief quotations in a book review.

Conatacts-www.topsmartphonesof2020.com

144 years ago ......

"Watson! If you can hear me, then go to the window and wave your hat."

- Alexander Graham Bell

(British Inventor, Scientist, Engineer, Professor)

What was in the past?

TOP SMARTPHONES OF 2020

What will be in the future?

Contents

Introduction

Smartphone is an important invention in the life of mankind. Development does not stand still. Now it's hard to surprise someone with new technology, but a few years ago the average person couldn't even dream of such an invention as a smartphone. Your smartphone allows you to always be in touch with the rest of the world, share the latest news with friends and inform your parents that everything is fine with you. Simply saying smartphones improve our lives each day. Now there is no need to be attached to a certain place, if there is a desire to talk with another person. This device allows you to call anywhere in the world.

Nowadays top smartphones could be not only a means of communication, but also a means of leisure.

On your device you can play games, watch movies, write a message, use the Internet and chat on social networks. It is not luxury anymore, so even a student and a pensioner have such a device.

Each person buys a smartphone according to his or her desire, appearance, and the necessary set of functions. It gives a lot of fantastic pleasure to an owner of a smartphone, as you are always in touch, you can always have fun when you are bored.

But have you ever thought, how much can smartphones be useful or harmful? Do you use them effectively? Do you know what your smartphone is really capable of? Are you planning to buy a new smartphone in 2020? And don't know between which to choose?

I guarantee some answers will be found in this book! I wish you a pleasant reading! ☺

THE MOMENT IT ALL STARTED

At the very beginning, people created mobile phones that had broader functionality, they were called communicators. They can truly be considered the ancestors of your current smartphones. Such communicators combined the functions of a telephone and a *PDA. At the beginning of the 21st century, these very communicators began to be called smartphones.

The term *"smartphone"* was first used by the manufacturer of mobile phones by Sony Ericsson in 2000. The company of developing mobile gadgets decided to emphasize the special features of the new Ericsson R380s. This model was the first smartphone in the world. It cost around US$700. Model had a touch screen, closed by a hinged lid and relatively small dimensions *(26 × 50 × 130 mm)* and weight *(170 g)*. Although it was called a smartphone, but still did not have the peculiarity of this device features - an open operating system.

----

*PDA- a Personal Digital Assistant, a.k.a. a handheld PC.

Later, in 2001 Nokia has released the first full-featured smartphone. He worked under the control of the open operating system Symbian OS version 6. 1.. In 2002, a whole line of smartphones was released. The most famous smartphone companies began to appear, one after another. Some of these companies were: Taiwanese company HTC; in the same year, the world-famous BlackBerry smartphones were born. They focused on working with e-mail, giving the user more opportunities.

When Microsoft released the operating system for Windows Mobile in 2003, smartphone production has begun gaining new momentum. Although Nokia remains true to Symbian OS, many manufacturers started launching Windows Mobile smartphones.

By 2006, the smartphone production market is approximately doubled. In the same year, Taiwanese electronic manufacturer HTC released the world's first 3G-smartphone based on the Windows Mobile operating system. Nokia is equipping its smartphones with functions previously characteristic only for communicators - Wi-Fi and GPS.

Even though by 2007 the concept of a smartphone had already thoroughly settled in the minds of users of mobile gadgets and it seemed that there was nothing more to come up with, the American company Apple makes a splash in this area by releasing the iPhone in 2007. The first iPhone model did not differ in particular functionality (there was no Bluetooth and the ability to send MMS messages), but the new way of controlling two-finger touch screen (Multi-Touch) and kinetic scrolling of images on the screen made it an icon in the world of mobile devices. For many, the iPhone has become an example of an ideal smartphone.

By the way, despite its iPhone ideality, as a matter of fact, all other electronic gadgets sometimes break. But the difference is that iPhone repair is only possible for highly qualified specialists - high-tech devices require the same attitude. Do not cost to ruin a device at the price of several hundred dollars to save several hundred dollars for its repair.

THE BEST ONES OF THE BEST ONES

Why rich people are increasingly buying very cheap phones?

Imagine! Ex manager of one of the world's largest photobanks Shutterstock, occupying two floors in the Empire State Building, 35-year-old Danny Groner rides the subway. He is the only one in the whole metro carriage who does not sit staring at the smartphone. Danny has a regular old-style push-button telephone, with which you can only call and send SMS, and he is proud of it. "I don't want to burn out," he explains. "I already spend 13-14 hours at work in front of the display, and there's no use sitting in front of it for 17 hours." At the same time, Groner observes that if everyone did the same, then work in the companies would have risen.

In the "telephone downshifting" billionaire Warren Buffett, the head of the financial giant Blackstone Stephen Schwartzman, actress Scarlett Johansson, singer Rihanna and other people who can afford the smartphones of the latest models have already been noticed. People who are tired of spending all their time on gadgets consciously choose independence from them.

Harvard University professor of psychology, Dr. Holly Parker, hopes the new habit will help people begin to share home and work. "This is a choice between the positions" my work remains for working hours" and "I am ready to work days and nights," she says. Parker adds that many companies would only benefit if they respected the right of their employees to rest, and did not ask them to be constantly in touch (this requirement is not in employment contracts, but de facto, as a rule, it is mandatory for office managers many large companies). Some even believe that it is time to spell out the law of a person's right to turn off his work phone. And the first country to do this already exists in the world - France.

"When tech-savvy, active people choose regular push-button telephones, it becomes a symbol of their status: they choose to save time and control themselves," says David Ryan Polgar, lawyer and technology ethicist. He notes that there are other ways to maintain this control - you just need to have the willpower to keep the smartphone away when you don't need it. At the same time, Groner admits that this method is not for him: "I do not trust myself. If I just decided to carry the switched off smartphone in my pocket, I would very soon break loose and again fall into this dependence."

Elin Shuk, head of the HR department at Accenture, an international consulting company, found a compromise: she has a regular push-button telephone "for home" and a smartphone for working hours. This is her second attempt to regain her free time. The first - when she promised herself to keep the iPhone away from the bedside table - failed. The second method worked better: now she admits that she no longer checks the "last time for today" mail and social networks before going to bed, and on the weekend she can lie on the beach without a smartphone.

How people of different countries use their phones?

Mobile phones are common in almost all countries of the world. However, residents of different countries, depending on traditions, culture and of course the level of wealth use smartphones in different ways. One analytics agency in 2019 surveyed residents of several countries and compiled detailed statistics on how they do this. The research involved several different countries from around the world, they were: USA, Australia, Russia, South Korea, China, Turkey, India, Brazil, the United Kingdom and Italy.

Researchers divided all mobile phones into 3 categories - smartphones (devices running on an "advanced" operating system), multimedia phones (devices with a touchscreen or QWERTY keyboard, but without a modern OS*) and feature phones (simple dialers).

---

*OS- Operating system (for ex.: Windows, IOS, Android).

It is not surprising that in developed countries, smartphones were much more popular than in developing countries. In UK, smartphones make 71% of all phones, in the South Korea it is 75%, the same percent in Australia. The back side can be observed in relatively poor countries. In India, feature phones make 67% of all mobile phones, in Turkey 51%, in Russia - 52%.

Multimedia Phones are not very popular. Advanced users prefer to immediately buy a smartphone, while ordinary users prefer not to spend too much. Only in Brazil and Turkey, these devices managed to get more than 25% of the market, in other countries their popularity is on average at 10%.

In different countries, the penetration rate of mobile phones is different, and this is not clearly related to the standard of living. For example, in Italy, 98% of the population over 16 years of age have a telephone with approximately the same number of mobile users as in the UK.

In most countries, with the exception of Australia, smartphone users are mostly men. Most of all, young people are focused on smart phones, which means that the smartphone market has great potential for growth. However, in Italy, people between the ages of 35 and 64 are most likely to use smartphones.

Advertising

About 65% of smartphone owners in Italy receive ads once a day or more often, in South Korea this figure is 82%, in Turkey - 79%, in China - 75%. So, it means that users from developed countries, see ads a little less often, and users from developing countries more often. However, there is an exception - the average smartphone owner from India sees ads no more than once a week.

Many people are willing to share personal information in order to receive precisely targeted advertising, for example, 42% of residents of China and Russia, 54% of Indians, 46% of Brazilians, 33% of Italians.

Video

Phone screen sizes, resolution, and overall quality continue to grow. So that is why mobile video is becoming increasingly popular.

Phone users are not inclined to believe that watching video from the phone affects their television preferences. However, in developing countries there is a deviation from this trend. About half of users from China said that they began to watch more television thanks to a smartphone. At the same time, 32% of the inhabitants of India, said that they began to watch less television under the influence of mobile video.

Vary the way to view mobile video. For example, 63% of the British and 58% of Koreans watch videos through applications. In the US, users equally often watch videos through a browser and through special applications. But 64% of users from India prefer to download video clips to the device and only then watch them.

Most often, mobile videos are watched in China and Brazil. About 19% of Chinese and 13% of Brazilians watch video clips from a smartphone several times a day. In developed countries we may see another picture, mobile video is less popular - about a third of users from the UK, Australia and Italy watch videos on the phone less than once a week. The exception is Americans, 33% of US residents watch videos on their phones once a day or more.

Specific Use and Acquisition

Many mobile phone owners prefer to have multiple devices to work with different mobile operators and to separate work and personal calls. In Russia, 55% of users have two or more mobile phones. In Brazil, 43% of people have several phones, in China 39%.

In developed countries, people prefer to have one phone: for example, in Italy several mobile phones have 37% of users, in Australia 26%, in the USA - 19%, in the UK and South Korea 18% each.

Depending on the country, people change the criteria for choosing a phone. For example, 35% of Americans, 32% of Italians and British, buy a phone with the best price. In Russia, phones are bought mainly on the basis of their appearance.

60% of Russians buy phones in large retail stores; 56% of Koreans prefer to buy devices from operators; 42% of Britons buy mobile phones on the Internet; Chinese, not surprisingly, take mobile phones directly from the manufacturer.

This very list of the best ones

The presented list of top smartphones of 2020 is objective at the time this book was written. Top smartphones choosing system was based on technological solutions of each of the gadgets, value for money, as well as their innovativeness.

Before you will see the list, it should be noted that top smartphones usually divide into three price classes:

❖ ENTRY- LEVEL (0$ - 200$)
❖ MID - RANGE (200$ - 600$)
❖ PREMIUM (600$ - 1500$)

Attention! Your opinion may slightly vary.

Let's get started...

10

Xiaomi Redmi Note 7

Main specifications:

- ❖ Camera - *48 Mpx; f/1.8*
- ❖ Processor - *Snapdragon 660*
- ❖ Memory - *32 gb*
- ❖ Screen - *LED IPS, Gorilla Glass 5*
- ❖ Battery - *4000 mAh*
- ❖ Weight - *186 g*
- ❖ 3,5 mm stereo audio jack - *Yes*
- ❖ Average price - *169 $*

Moto G7 Play

Main specifications:

- ❖ Camera - *13 Mpx; f/2.0*
- ❖ Processor - *Snapdragon 632*
- ❖ Memory - *32 gb*
- ❖ Screen - *HD+ Max Vision display*
- ❖ Battery - *3000 mAh*
- ❖ Weight - *149 g*
- ❖ 3,5 mm stereo audio jack - *Yes*
- ❖ Average price - *145 $*

Samsung Galaxy A40

Main specifications:

- ❖ Camera - *16 Mpx; f/2.0.*
- ❖ Processor - *Exynos 7904*
- ❖ Memory - *64 gb*
- ❖ Screen - *Infinity-U display*
- ❖ Battery - *3100 mAh*
- ❖ Weight - *140 g*
- ❖ 3,5 mm stereo audio jack - *Yes*
- ❖ Average price - *199 $*

Moto G6

Main specifications:

- ❖ Camera - *12 Mpx; f/1.8; f/2.55"*.
- ❖ Processor - *1.8 GHz octa-core*
- ❖ Memory - *32/64 gb*
- ❖ Screen - *Full HD+ display*
- ❖ Battery - *3000 mAh*
- ❖ Weight - *167 g*
- ❖ 3,5 mm stereo audio jack - *Yes*
- ❖ Average price - *195 $*

5

Huawei Y7

Main specifications:

- ❖ Camera - *13 Mpx; f/1.8*
- ❖ Processor - *Snapdragon 450*
- ❖ Memory - *32 gb*
- ❖ Screen - *LED IPS*
- ❖ Battery - *4000 mAh*
- ❖ Weight - *168 g*
- ❖ 3,5 mm stereo audio jack - *Yes*
- ❖ Average price – *175 $*

4

Xiaomi Mi A2

Main specifications:

- ❖ Camera - *20+12 Mpx; f/1.8;f/2.8.*
- ❖ Processor - *Snapdragon 660 AIE*
- ❖ Memory - *32 gb*
- ❖ Screen - *IPS LCD*
- ❖ Battery - *3010 mAh*
- ❖ Weight - *168 g*
- ❖ 3,5 mm stereo audio jack - *No*
- ❖ Average price - *180 $*

3

Huawei P Smart

Main specifications:

- ❖ Camera – *13+2 Mpx; f/1.8.*
- ❖ Processor - *Hisilicon Kirin 659*
- ❖ Memory - *32 gb*
- ❖ Screen - *Infinity-U display*
- ❖ Battery - *3400 mAh*
- ❖ Weight - *143 g*
- ❖ 3,5 mm stereo audio jack - *Yes*
- ❖ Average price - *180 $*

2

Honor 10 Lite

Main specifications:

- ❖ Camera - *13 Mpx; f/1.8.*
- ❖ Processor - *4 x 2,2 GHz + 4 x 1,7 GHz*
- ❖ Memory - *64 gb*
- ❖ Screen - *IPS LCD*
- ❖ Battery - *3400 mAh*
- ❖ Weight - *162 g*
- ❖ 3,5 mm stereo audio jack - *Yes*
- ❖ Average price - *199 $*

1

Motorola One

Main specifications:

❖ Camera – *13+2 Mpx; f/2.0, 1/3.1"*.

❖ Processor – *Snapdragon 625*

❖ Memory - *64 gb*

❖ Screen - *HD+ display*

❖ Battery - *3000 mAh*

❖ Weight - *162 g*

❖ 3,5 mm stereo audio jack - *Yes*

❖ Average price – *170 $*

10

Xiaomi Note 10

Main specifications:

- ❖ Camera - *108 Mpx; f/1.7,1/1.33".*
- ❖ Processor - *Kryo 470 Silver*
- ❖ Memory – *128 gb*
- ❖ Screen - *AMOLED, Gorilla Glass 5*
- ❖ Battery - *5260 mAh*
- ❖ Weight - *208 g*
- ❖ 3,5 mm stereo audio jack - *Yes*
- ❖ Average price – *560 $*

Samsung Galaxy A51

Main specifications:

- ❖ Camera - *48 Mpx; f/2.0; f/2.2.*
- ❖ Processor – *2.3 GHz*
- ❖ Memory – *128 gb*
- ❖ Screen - *Full HD+ Super AMOLED*
- ❖ Battery - *4000 mAh*
- ❖ Weight - *172 g*
- ❖ 3,5 mm stereo audio jack - *Yes*
- ❖ Average price – *399 $*

iPhone 8

Main specifications:

- ❖ Camera - *12 Mpx; f/1.8.*
- ❖ Processor - *Apple A11 Bionic*
- ❖ Memory – *64 gb*
- ❖ Screen - *ED IPS, 3D Touch*
- ❖ Battery - *1821 mAh*
- ❖ Weight - *148 g*
- ❖ 3,5 mm stereo audio jack - *No*
- ❖ Average price – *549 $*

Samsung Galaxy S10e

Main specifications:

- ❖ Camera - *12 Mpx; f/1.5-2.4.*
- ❖ Processor - *Exynos 7904*
- ❖ Memory – *128 gb*
- ❖ Screen – *Super AMOLED*
- ❖ Battery - *3100 mAh*
- ❖ Weight - *175 g*
- ❖ 3,5 mm stereo audio jack - *Yes*
- ❖ Average price – *539 $*

Huawei P30

Main specifications:

- ❖ Camera - *40 Mpx; f/1.8.*
- ❖ Processor – *Kirin 980*
- ❖ Memory – *128 gb*
- ❖ Screen - *OLED*
- ❖ Battery - *3650 mAh*
- ❖ Weight - *165 g*
- ❖ 3,5 mm stereo audio jack - *Yes*
- ❖ Average price – *199 $*

5

Xiaomi Redmi Note 8 Pro

Main specifications:

- ❖ Camera - *64 Mpx; f/1.9.*
- ❖ Processor - *Mediatek Helio G90T*
- ❖ Memory – *128 gb*
- ❖ Screen - *IPS LCD, Gorilla Glass 5*
- ❖ Battery - *4500 mAh*
- ❖ Weight - *199 g*
- ❖ 3,5 mm stereo audio jack - *Yes*
- ❖ Average price – *250 $*

4

Xiaomi Mi 9T Pro

Main specifications:

- ❖ Camera - *48 Mpx; f/1.8.*
- ❖ Processor – *Snapdragon 855*
- ❖ Memory – *128 gb*
- ❖ Screen - *AMOLED, Gorilla Glass 5*
- ❖ Battery - *4000 mAh*
- ❖ Weight - *191 g*
- ❖ 3,5 mm stereo audio jack - *Yes*
- ❖ Average price – *480 $*

Samsung Galaxy A70

Main specifications:

- ❖ Camera - 32 *Mp; f/1.7.*
- ❖ Processor – *Snapdragon 675*
- ❖ Memory –128 *gb*
- ❖ Screen - *Super AMOLED*
- ❖ Battery - *4500 mAh*
- ❖ Weight - *183 g*
- ❖ 3,5 mm stereo audio jack - *Yes*
- ❖ Average price – *420 $*

2

Samsung Galaxy A40

Main specifications:

- ❖ Camera - *12 Mpx; f/1.7.*
- ❖ Processor - *Snapdragon 636*
- ❖ Memory – *64 gb*
- ❖ Screen - *Gorilla Glass 5*
- ❖ Battery - *3000 mAh*
- ❖ Weight - *180 g*
- ❖ 3,5 mm stereo audio jack - *Yes*
- ❖ Average price – *450 $*

Samsung Galaxy A80

Main specifications:

- ❖ Camera - *48 Mpx; f/2.0..*
- ❖ Processor - *Snapdragon 730*
- ❖ Memory – *128 gb*
- ❖ Screen - *Super AMOLED*
- ❖ Battery - *3700 mAh*
- ❖ Weight - *219 g*
- ❖ 3,5 mm stereo audio jack - *Yes*
- ❖ Average price – *480 $*

10

iPhone 11 Pro Max

Main specifications:

- ❖ Camera - *12 Mpx; f/1.6-2.2.*
- ❖ Processor – *A13 Bionic*
- ❖ Memory – *256 gb*
- ❖ Screen - *OLED, Super Retina XDR*
- ❖ Battery - *3500 mAh*
- ❖ Weight - *226 g*
- ❖ 3,5 mm stereo audio jack - *No*
- ❖ Average price – *1500 $*

Samsung Galaxy S10

Main specifications:

- ❖ Camera - *16 Mpx; f/1.5-2.4.*
- ❖ Processor - *Exynos 9820*
- ❖ Memory – *512 gb*
- ❖ Screen – *Super AMOLED*
- ❖ Battery - *3400 mAh*
- ❖ Weight - *157 g*
- ❖ 3,5 mm stereo audio jack - *Yes*
- ❖ Average price – *1050 $*

Samsung Galaxy Note 10

Main specifications:

- ❖ Camera - *16 Mpx; f/1.7.*
- ❖ Processor - *Exynos 9825*
- ❖ Memory – *256 gb*
- ❖ Screen - *Dynamic AMOLED*
- ❖ Battery - *3500 mAh*
- ❖ Weight - *168 g*
- ❖ 3,5 mm stereo audio jack - *No*
- ❖ Average price – *899 $*

iPhone XS

Main specifications:

- ❖ Camera - *12 Mpx; f/1.8.*
- ❖ Processor – *A12 Bionic*
- ❖ Memory – *64 gb*
- ❖ Screen – *3D Touch, OLED*
- ❖ Battery - *3174 mAh*
- ❖ Weight – *177 g*
- ❖ 3,5 mm stereo audio jack - *No*
- ❖ Average price – *899 $*

Sony Xperia 5

Main specifications:

- ❖ Camera - *12 Mpx; f/1.6.*
- ❖ Processor – *Snapdragon 855*
- ❖ Memory – *128 gb*
- ❖ Screen - *OLED, Gorilla Glass 6*
- ❖ Battery - *3140 mAh*
- ❖ Weight - *164 g*
- ❖ 3,5 mm stereo audio jack - *No*
- ❖ Average price – *899 $*

5

Sony Xperia 1

Main specifications:

- ❖ Camera - *12 Mpx; f/1.6.*
- ❖ Processor – *Snapdragon 855*
- ❖ Memory – *128 gb*
- ❖ Screen - *OLED, Gorilla Glass 6*
- ❖ Battery - *3330 mAh*
- ❖ Weight - *180 g*
- ❖ 3,5 mm stereo audio jack - *no*
- ❖ Average price – *950 $*

4

Samsung Galaxy s10 Plus

Main specifications:

- ❖ Camera - *16 Mpx; f/1.5-2.4.*
- ❖ Processor - *Exynos 9820*
- ❖ Memory – *128 gb*
- ❖ Screen – *QHD+*
- ❖ Battery - *4100 mAh*
- ❖ Weight - *175 g*
- ❖ 3,5 mm stereo audio jack - *Yes*
- ❖ Average price – *999 $*

Huawei P30 PRO

Main specifications:

- ❖ Camera - *32 Mpx; f/1.6.*
- ❖ Processor – *Kirin 980*
- ❖ Memory – *64 gb*
- ❖ Screen - *OLED*
- ❖ Battery - *4200 mAh*
- ❖ Weight - *189 g*
- ❖ 3,5 mm stereo audio jack - *No*
- ❖ Average price – *699 $*

OnePlus 7T PRO

Main specifications:

- ❖ Camera - *48 Mpx; f/1.6.*
- ❖ Processor – *Snapdragon 855*
- ❖ Memory – *64 gb*
- ❖ Screen - *Corning Gorilla Glass 3D*
- ❖ Battery - *4085 mAh*
- ❖ Weight - *206 g*
- ❖ 3,5 mm stereo audio jack - *Yes*
- ❖ Average price – *699 $*

1

Google Pixel 4

Main specifications:

- ❖ Camera - *16 Mpx; f/1.4.*
- ❖ Processor – *Snapdragon 855*
- ❖ Memory – *64 gb*
- ❖ Screen – *P-OLED*
- ❖ Battery - *2800 mAh*
- ❖ Weight - *162 g*
- ❖ 3,5 mm stereo audio jack - *Yes*
- ❖ Average price – *679 $*

WHAT SHOULD BE THE BEST CAMERA PHONE?

Today, many smartphone users can say- „Making amazing shots no worse than on a DSLR camera is quite affordable for smartphones with a good camera!". But what do users mean by saying- „good"?

If you ask them this simple question, in most cases you will hear something like- „I bought this phone because it had more megamixels and cameras than others!". But is it so? Do megapixels even matter? I do not think so. To those people who did not know this fact, I will explain.

Megapixels mean nothing

Imagine an ordinary window in your house. The number of megapixels, is the number of glasses inside the window frame. If you continue to draw parallels with smartphones then, in ancient times, glass for windows were the same size and were considered a scarce commodity.

Therefore, when somebody could tell you that he had 7 glasses (megapixels) in his window, then everyone understood that this person was a serious and wealthy man. And the characteristics of the window were also immediately clear - a good view outside the house, a large area of glazing.

A few years later, windows (megapixels) stopped to be a deficit, so their number had only to be brought to the required level, and calm down. It's easy to bring it into line with the area (a window for ventilation and a loggia, for the sake of strength, they require a different number of windows) so that the camera produces a slightly denser picture than 4K monitors and TVs produce.

And finally to deal with other characteristics - for example, to combat the clouding of glass and image distortion. To teach cameras to sharpen and paint existing megapixels properly, if you want specifics.

It is all about aperture

Aperture is needed to measure a certain amount of light. The more it expands, the more light falls on the photo module matrix. This characteristic of a smart camera is measured in f-stops. The size range consists of fractional numbers, for example, f / 1.4. The larger the number on the right, the less the aperture is revealed. For example, in the iPhone 11, the main photo module has indicators of 1.8 and 2.4.

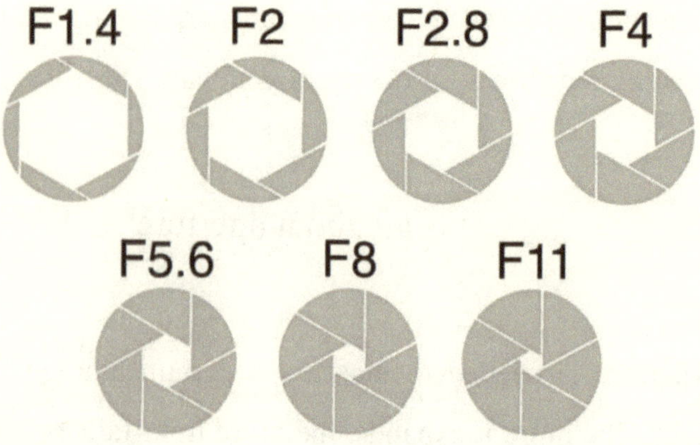

How can all these numbers affect the picture? A large aperture transmits the maximum amount of light to the sensor, while a small aperture passes the minimum. In poor light conditions, you need to maximize the opening of the lens shutter, but in bright light, on the contrary, you can thus illuminate the frame. Small aperture values will be useful to fans of portraits and landscapes, because this increases the sharpness of the picture.

Nowadays, even some budget phones can boast the size of the rear photo module in f / 2.2. If for evening shooting with devices with cameras 2.4 or 2.6 you must resort to use additional applications to achieve photo brightness, then units with f / 2.2 and less will take a good picture without additional effort. On average, it is better to focus on indicators - f / 1.8-1.6. For example, in a P20 Pro with a triple main photo module, this value is 1.6; 1.8; 2.4.

The aperture itself and its performance do not guarantee maximum photo quality. But in conjunction with other parameters (which will be discussed below), it is better to choose smartphones in which cameras do not "squint" in order to ultimately get frames that are sufficiently illuminated and clear. Or you can pay attention to smartphones such as Samsung S9, which is equipped with a "smart" aperture shift from 2.4 to 1.5 depending on weather and lighting.

NEAR FUTURE OF SMARTPHONES

Charging by air

This is not about wireless charging stations, if someone thought so, but about a complete separation from the outlet. Imagine the transmitter towers everywhere, constantly charging your phone and keeping the battery at maximum. You will not to have wires and power banks anymore!

Utopian? I do not think so… The American company Ossia in January of 2019 introduced the concept of a device that is capable of generating a charging current at a distance of up to one meter, and a pair of such transmitters can recharge a smartphone at a distance of up to nine meters. A gadget called Cota will still be running in Walmart's warehouses, recharging trackers to track inventory.

Hybrid screen

OLED displays are good for games and watching videos, but reading from them is not very good. E-books with E-Ink technology are much more pleasing to the eye. And although on the LED screen you can play with brightness and contrast, anyway, it can not be compared with "electronic ink".

Future smartphones will probably kill readers (or swallow them - as you like). And no, it will not be like the Yotaphone 3 with two screens, but a full-fledged hybrid screen. Apple had a similar idea - a combined LCD / E-Ink display that switches depending on what is currently displayed on the screen. 8 years ago, the idea was not brought to mind and commercial use - but it would be worth it.

Real time translation

The world is rapidly globalizing, and all (well, almost all) countries of the world can be seen with their own eyes. For travel it is worth knowing the English language - and most of our readers probably know it at a level sufficient to find out the way to the station. But in many countries, most locals are not friendly with English, and we are not talking about third world countries - Spain is an example. And with those who seem to know, sometimes I want to talk more thoroughly, and my vocabulary limits it.

Here it's worth recalling why smartphones were created initially - not for surfing the Internet, but for communication. Although Google Translate is still experiencing some problems with the translation of written texts, the development of neural networks should help smartphones recognize and understand "by ear" text in another language, as well as translate it to the owner.

Color changing case

The same color is boring, and constantly changing covers is somehow silly, and I don't want to spend extra money. Surely in the future, phone manufacturers will figure out how to change the color at the touch of a button (or with one effort of thought, since we are seriously fantasizing here).

The easiest way to implement such a mechanism: a housing made of glass-like material with built-in LEDs. So you can change the color of the phone at least once a second, and if the LEDs are positioned correctly, then make it gradient. And of course I would like this glass-like material to be unbreakable.

Mental management

Scientists are approaching reading thoughts step by step, and very soon they will be able not only to fix them in the form of teeth on an echoencephalogram, but to decode them in an understandable form. And engineers will immediately try to create a mental interface on this basis.

Type text, change brightness, find the file you need - all this can be done without using voice commands or swipe with your finger, you just have to think carefully. Although there are still not a year or two work up to such opportunities, there are already certain advances in the direction of mentally controlling gadgets. Facebook, for example, has been working on technology for mental typing for two years now.

STATISTICS & FACTS

1. More than a billion smartphones were released in 2019, and more than 867 million units were released in the first half of 2018.

2. The most expensive elements of a smartphone are the screen and memory.

3. In South Korea, a "disease" from a smartphone was invented - digital dementia. It has been proven that if you get carried away using a smartphone, then the person loses the ability to concentrate.

4. More than 20 billion applications are downloaded annually to smartphones.

5. Today, there are more smartphones than toilets in India.

6. Fins have created a new sport - throwing smartphones. This is due to the fact that they are tired of fighting addiction to modern gadgets.

7. Japanese use a smartphone even while taking a shower.

8. German Chancellor Angela Merkel has 2 smartphones.

9. At the heart of each smartphone is the operating system.

10. People when buying a smartphone today pay more attention not to hardware, but to the device's software.

11. The term "smartphone" was introduced by Ericsson in 2000 to designate its own new Ericsson R380s phone.

12. The price of the first smartphone was approximately $ 900.

13. Literally, "smartphone" translates as "smart phone".

14. A smartphone has much greater computing power than a computer that delivers astronauts to the moon.

15. Nomophobia is the fear of being left without a smartphone.

16. More than 250 thousand patents are based on smartphone technology.

17. Approximately 110 times daily, an average person looks at his smartphone.

18. Most smartphones in Japan are waterproof.

19. About 65% of smartphone users do not download applications to it.

20. Approximately 47% of Americans could not live a day without using a smartphone.

21. The first smartphone was a commercial touch device that can be controlled with a stylus and the usual touch of fingers.

22. Modern smartphones are "gluttonous" devices.

23. The very first thin smartphone is considered to be a gadget made in Korea. Its thickness was only 6.9 millimeters.

24. The weight of the world's first smartphone was only 400 grams.

25. A disorder in which a person is afraid to answer calls on a smartphone is called telephobia.

26. Total 2 types of the most expensive smartphones exist in the world. This is a Vertu gadget and custom iPhone

27. About 1140 calls per year are made from a smartphone.

28. The world's first smartphone was released 20 years after the first mobile phone.

29. Each 3rd smartphone holder consults friends before buying it.

30. About 18 thousand messages are sent every year by a teenager in Great Britain.

References

1. Simon, K. (2019). Digital 2019: Global digital overview. *Slides 1-221.* (https://datareportal.com/reports/digital-2019-global-digital-overview).

9 798604 733363